GW00457187

FAMOUS LAST LINES

This edition first published in 1993 by
Malcolm Holdsworth, George Bambery,
Robert Kingsford-Smith.

Production and design by
Goanna Graphics with special thanks
to Jacqueline Ballard.

National Library of Australia
Cataloguing-in-Publication data:

Famous Last Lines

ISBN 0 646 14178 3.
1. Locomotives. 2. Railroad travel.
I. Holdsworth, Malcolm.
II. Bambery, G.
III. Kingsford-Smith, R.

625.261
Produced in Hong Kong
by Mandarin Offset.

The serried ranks of the Grooswartberge dominate this early morning view of 19D 3340 near Stolsvlakte with a Port Elizabeth to Cape Town passenger. April 1979 MT

Famous Last LINES

COMPILED BY MALCOLM HOLDSWORTH

CONTENTS

Introduction

Steam locomotives have been objectively examined for generations. Their dimensions and performance have been measured, weighed against those of engines from directly competing railway systems and, in the case of superlative designs, even compared with feats of strength and speed on a global basis.

The moods of steam have captured the fancy of man for generations, affecting each observer in an intimate manner and drawing evocative tales from the pens of many of the world's great writers.

Steam locomotives have been photographed for as long as photography has existed. Some photographers have developed a sensitivity for the subject matter which has become legendary.

The best of their images must rate as high art and their published works endure as a memorial to the steam age.

This is a hard act to follow!

Famous Last Lines was conceived with the premise that world steam operation is far too large a topic to cover in one book. The decision was taken to select five lines of diverse character and to present a written and photographic essay on each. In doing so, we have leant heavily towards the photographic.

In their youth, most of the contributors to the book were exposed to the bedrock experience of mainline steam action on the Short North line in New South Wales, Australia. The source of a lifetime's infatuation could not be denied its place between the covers!

Debate then raged on the merits of other favourite haunts around the globe, with the twin considerations of scenic appeal and uniqueness of operation uppermost in our minds. Focus finally sharpened on railway activity in the wilds of Patagonia, the snows of Vordernberg, the coast of Cape Province and the heights of Darjeeling.

In attempting to capture the diverse nature of these operations, we contemplated the variety of landforms, weather patterns, language, locomotive types and railway practice. In looking at all these differences, a common factor emerged. It was the unfailing good humour and camaraderie of every country's railwaymen, both in their dealings with each other and with ourselves. When one considers the harsh working conditions associated with steam, a fireman's ability to share a joke or a cup of tea after shovelling tons of coal says a lot.

With this positive response from railway staff in mind, it is easy to understand why so many enthuasists dedicate their publications to the people who make the trains go. We acknowledge their support with equal sincerity.

We also dedicate this book to the hoteliers, mechanics, travel agents and hire car drivers who supplied the other side of the equation — feeding us and getting us to our photographic locations. Awakened often in the small hours; exhorted to repair the irreparable NOW!; requested to book flights to places unknown and instructed to drive FASTER!: they are the unsung achievers behind the people behind the cameras. That firm friendships have grown over the years between many of these stalwarts and their zealous, fixated clients is a continuing source of wonder.

Famous Last Lines is thus a conglomerate of effort and art. Whether the artistic outcome we have sought has been realised is an issue which readers must decide for themselves. But we assure you of one thing. This production has been a labour of love.

THE SHORT NORTH

THE GOSFORD
— NEWCASTLE
RAILWAY,
NEW SOUTH WALES

AUSTRALIA

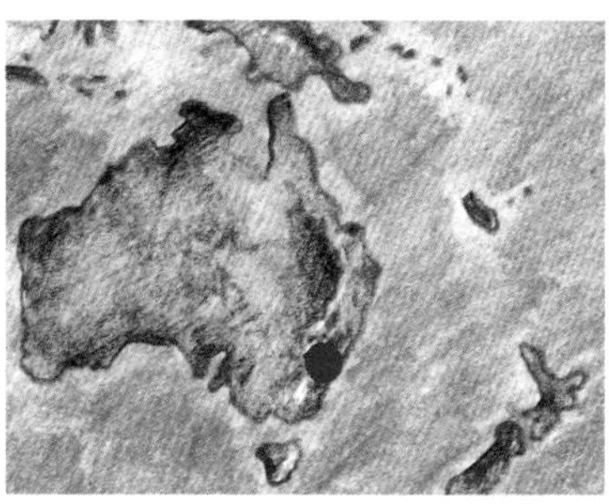

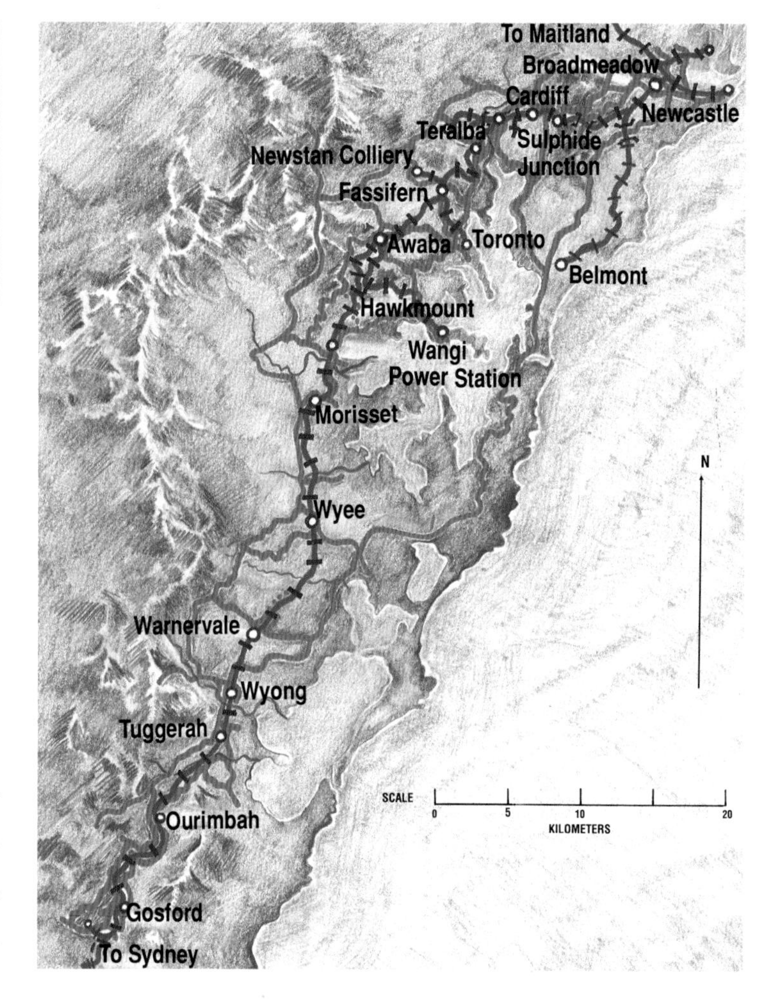

STATUS:	Two Track Main Line
GAUGE:	4' 8 1/2" (143.5cm)
LENGTH OF LINE:	53 Miles (85 Kilometres)
RULING GRADES:	1 in 40 For Newcastle Bound Trains 1 in 75 For Gosford Bound Trains
TRAFFIC:	Express and Local Passenger Services General Freight Services
MOTIVE POWER:	30 Class 4-6-4 Tank 1903 Beyer Peacock* 30T Class 4-6-0 1928 Eveleigh*# 35 Class 4-6-0 1914 Eveleigh 36 Class 4-6-0 1925 Eveleigh* 38 Class 4-6-2 1943 Clyde* 50 Class 2-8-0 1896 Beyer Peacock* 53 Class 2-8-0 1912 Clyde* 59 Class 2-8-2 1952 Baldwin 60 Class 4-8-4+4-8-4 1952 Beyer Peacock * First Builder # Modified from 30 Class

A ride on 19 North

School finished at about 3.30 on Friday afternoon and with a feeling of mounting excitement I raced home, pausing at the chemist's for a roll of Kodachrome. One roll was all I could afford and was far and away the weekend's biggest purchase.

Because the nights were cold, a spare pullover was packed with Mum's sandwiches, a couple of cans of drink, chocolate bars, matches and the Dolphin torch. The surging adrenalin subsided after dinner to allow a few hours' sleep, followed by the half-drugged sensation of waking to the alarm at midnight. A long suffering father collected those of my mates who were also going that night and deposited us at Hornsby Station to await the "Paper Train", No. 19 North.

Electrification having come to the 50 mile section of the Short North from the New South Wales capital to Gosford in 1960, our journey on the 1.00am ex Sydney commenced behind a Metropolitan Vickers built 46 Class Co+Co. The quiet efficiency of this form of traction was always an unsettling reminder of the advance of technology. As our "Beetle's" headlight swung around the curves of Mullet Creek, the echoing roar of Alco power on a preceding express goods climbing up to Woy Woy tunnel underlined the limit of time left to steam. Once through the tunnel in our turn, it seemed only minutes before we crossed the Broadwater, turned gently right then more sharply left into the platform at Gosford.

Many an anxious passenger has leant out of the windows of 19 North to get a first glimpse of Gosford Loco. Invariably there was steam, but would diesels be lurking in the yard as well? What would the Chargeman's Office at Broadmeadow have rostered to replace the departing electric? Fortunately for all concerned, more often than not in the late 60s the gleaming bulk of a 38 Class Pacific would emerge from the gloom and gently lock couplers with its load of bogie freight wagons, heavy passenger cars and mail vans. Such was the case on this occasion.

After considerable delay (19 was that sort of train) the "Eighter's" chime whistle sounded, the brakes were released and we drew slowly past the slumbering shapes of an ever-varied assortment of "small" engines — Standard Goods 2-8-0s, 35 and 36 Class 4-6-0s, 38 Class 4-6-2s and 59 Class 2-8-2s. While our sense of smell absorbed the odour of coal smoke, lubricating oil and hot metal with relish, it was also assailed by the reek of sheep in vans on the "up" side of the yard, waiting their turn at the meat works.

Picking up speed, we travelled under the heavy brick arch of the road bridge at the northern end of the yard and then passed the giant forms of 60 Class 4-8-4+4-8-4s in differing states of readiness at the Garratt stage. We did our best to identify the class and engine number of as many of the locomotives as possible, because they would be the participants in the procession of single and double headed combinations to turn up before dawn and into the morning at our chosen photo location.

The rest of the trip on 19 flew by in the exhilaration of stop-start train working, the big Pacific whistling forcefully at level crossings, marching out of dimly lit halts that never saw an express below 70 mph and rushing through the invisible countryside. Opposing movements passed with a hiss and a rumble; occasionally the moon uncovered our locomotive as silver steam and stinging cinders cascaded back along the train.

If we were alighting at a recognised station we had no problems. Descending at conditional halts required only a word to the guard. To get off at isolated locations like Hawkmount, roughly in the middle of the six mile stretch between Dora Creek and Awaba and at the top of a nasty 1 in 44 climb for "down" trains, an act of charity was needed on the part of the driver.

Surprisingly, among the hardy crews were a number of souls who not only didn't mind stopping in the bush in pitch dark conditions, but apparently relished the task of restarting over 300 tons of train on the worst part of the grade — a tight right hand curve about a half mile from the summit.

We jumped down onto the ballast with our gear, the guard made sure we were OK, then gave the driver the "right of way" with his lamp. A loud chime came from up front as if the train were leaving a scheduled stop, a couple of explosive slips as the Boxpok drivers lost traction on the dewy metal, then the crashing exhaust of well harnessed horsepower accelerating into the distance.

After setting up our camp, we grabbed old sacks and disused metal drums and descended to track level to collect lumps of coal, shed from the tenders of passing locomotives. With our fire well lit, we sat through the balance of the night, watching the pale orange beams of steam headlights sweeping the eucalypt forest. The soft exhaust beats of the old "Standards", the sharper stack talk of the younger "smalls" and the shuffling roar of the Garratts wove a spell of power over our little group. By shining our torches, we often brought a volley of whistles from the busy crews and as the engines slogged past our camp spot we rushed to the cutting's edge to glean the numbers from the cab sides.

With the dawn, the labouring engines came within range of a battery of photographic equipment and our early efforts must speak for themselves.

After Saturday's photography was over, we trudged to the nearest station, boarded a train, sat semi-comatose from an overdose of steam, got home, washed off a thick layer of soot and fell into bed for ten hours of dreamless sleep. But then why dream? The photos were in the bag, the memories couldn't have been excised by a skilled lobotomist and Friday night was only six days away! MH

Last light catches 6037 with over 600 tons behind the drawbar at the summit of the 1 in 40 grade north of Fassifern.
December 1970 CS

▲

Dawn on the second of November 1968 and a light engine Garratt drops towards Fassifern on its way to a day's coal haulage on the Wangi Branch.

November 1968 RB

Garratts converge near Wyee. The double header running towards the camera is facing the opposite way to normal practice as the turntable at Broadmeadow was under repair. May 1971 MH ▶

The "tortoise" in the form of ageing Consolidation No 5369 comes face to face with the "hare" as 3806 sprints an early morning passenger service up the 1 in 75 wooded northern side of Fassifern Bank.

August 1967 SMcC

Flyer! Non-streamlined 38 Class Pacific No 3809 gallops the 310 ton HUB set of No 24 Newcastle to Sydney express upgrade from Booragul one early July morning. July 1967 RB

A Nanny faces the dawn.
September 1967 SMcC

▲

The mechanical stoker on 6018 is turning as fast as it can and 63,000 lbs of tractive effort is being applied to rail on the long 1 in 44 grade of Hawkmount. June 1970 RKS

The top of a large stringybark tree provides an aerial vantage point for the exertions of 6037 on 662 steel train ex Port Waratah. The location is Hawkmount. July 1970 RKS ▶

▲
The marker lights of Mikado No 5920 struggle to pierce the fog at Sulphide Junction.
May 1972 RKS

At night, at rest. Cab interior of ▶
6039 at Gosford.
December 1971 MH

By 1972 a Standard Goods at work on the Short North was considered a thing of the past. However an oil shortage in May of that year brought a few of the veteran 2-8-0s back to Gosford for a final fling.
May 1972 MH

▲ The clocktower at Newcastle Terminal Station rises in silhouette behind a 30T 4-6-0 on a Maitland local and a 38 on the up morning Flyer.
March 1969 RB

Green Pacific No 3813 pauses under the timbers of the coal stage at Broadmeadow. ▶
October 1969 CS

▲ A bird's eye view of Broadmeadow Numbers 1 and 2 roundhouses. Number 1 on the left housed tender and tank engines during steam's last years. Number 2 held the Garratts and diesels.
October 1969 LW

◀ The prevailing colour in steam sheds the world around is grey. The confines of Broadmeadow No 2 roundhouse were no exception. April 1966 RB

◀ 6017 and a sister lift maximum tonnage beside the infamous "Hawkmount Hotel".
May 1968 RKS

The "Hawkmount Hotel" offered vantage points and accommodation extraordinary to the knowledgeable few. Several of the guests take the air between bursts of activity. ▼
September 1970 RKS

◀ The ultimate spectacle on the Short North was the contest between double 60 Class on trains out of Newstan Colliery and the 1 in 40 grade of Fassifern Bank. The Garratts would propel the often grossly overloaded trains backwards through the station to take advantage of the short falling grade, and then . . . attack! No matter how well they were doing at the bottom, walking pace was the norm at the summit. When first used on these duties, six diesels failed in one day with blown traction motors. December 1970 CS

With exhaust noise reverberating off the gum trees, a "Pig" (36 Class) and Garratt tramp upgrade from Fassifern late one afternoon.
June 1967 RB

For a few minutes the plodding exhaust of a Standard Goods on a full load had crawled closer. Finally train No 269, the pick up goods, hove into view behind 5485 and worked up the valley to Tickhole Tunnel on Anzac Day. April 1969 RB

3134 restarts a Toronto line passenger service from Teralba Station, Newcastle bound. August 1969 CS

Having trundled its consist across the tidal flats from Toronto, a 30 tank restarts its morning commuter train from the platform at Fassifern.

April 1969 LW

▲

The setting sun tints the area near Cockle Creek with gold as a Newcastle-bound passenger jogs by. The colour light "stick" on the left was a rarity, confined to the inner suburban area. April 1969 LW

◀ The Toronto line branched off at Fassifern for a water level run to the shores of Lake Macquarie. This was tranquil stuff. The return up the 1 in 40 towards Newcastle was not. May 1967 RB

3531 leads a Gosford local into the cutting at the summit of Fassifern Bank. June 1967 RB ▶

Light and steam — 3809 stretches its 5'9" legs near Ourimbah on the head end of the up morning Flyer.
August 1968 SMcC ▼

ESQUEL

THE ESQUEL BRANCH OF THE FERROCARRIL GENERAL ROCA SYSTEM, PATAGONIA

ARGENTINA

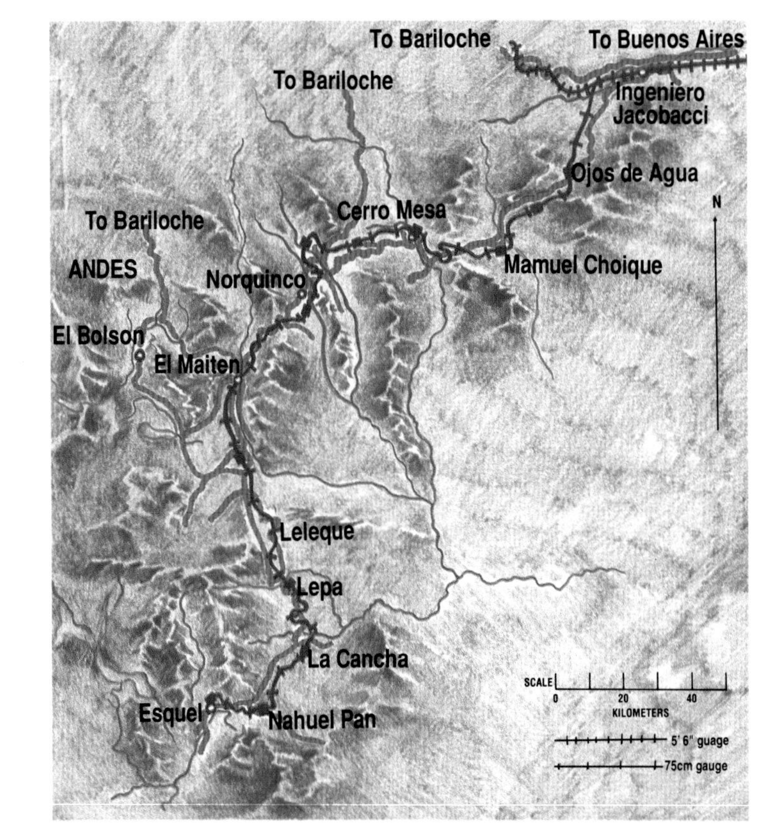

STATUS:	Single Track Branch Line
GAUGE:	2' 6" (75cm)
LENGTH OF LINE:	250 Miles (400 Kilometres)
TRAFFIC:	At Least One Return Passenger Service Per Week General Freight As Required
MOTIVE POWER:	75B Class 2-8-2 1922 Baldwin 75H Class 2-8-2 1922 Henschel 75M Class 0-6-0T 1922 Henschel

Cold days, hot engines

A famous travel writer set out from Boston one dreary winter's morning on a journey that would take him many thousands of kilometres to the south: to another continent and another hemisphere. He travelled by train, or rather by many trains, filling in some trackless gaps by bus or other unspeakable forms of transport. Months later his final train journey took him to his destination and provided the name for a book. It took him to Esquel, a town in an Andean valley in Patagonia, the wild, forgotten southern tip of South America.

The rail traveller to Esquel disembarks from a broad gauge (5'6") train at Ingeniero Jacobacci, a remote, unattractive little town at the western extremity of the Patagonian plain. On the other side of the low level platform the timbers of the narrow gauge (2'6") carriages groan as heavily clad transiting passengers climb aboard in the freezing pre-dawn. Two ancient Baldwin 2-8-2s back onto the train. Departure is signalled by a high pitched whistle tone strangely out of character with the North American origins of these locomotives.

After a few kilometres of dual track the broad gauge is left behind. The narrow gauge swings south-west to begin a journey of 400 lonely, tortuous kilometres. A stutterous clamour drifts back from the front of the train as the locos begin the assault on the first long climb. It is mid-winter and at isolated stops water column leakage has frozen in gleaming sheets, emphasising the dryness of the land. The Andes, away to the west, trap most of the moisture from the westerlies of the South Pacific, leaving this corner of Patagonia dry and treeless. A huge, flat topped mountain guards the cleft through which the line crosses the first of the great ridges which the Andes have thrust eastwards into the desert and which impede passage to the south.

Occasionally the line crosses a river. On the broad flats, homesteads (estancias) are losing a futile battle against winter gales, protected only by drought-stunted clumps of leafless foreign trees. Their isolation is a reminder that the pioneering days of European settlement in this land are not long past. Lost-looking sheep and jittery flocks of large flightless rheas reinforce the feeling of human intrusion into a vast, desolate wilderness. But the illusion is shattered when the train labours to the top of another long grade. As the clouds of oil smoke clear, the traveller catches a tantalising glimpse of the still distant Andes.

At one of the river crossings the train shuffles to a stop beside the hamlet of Cerro Mesa. As the locos replenish thirsty tenders with precious river water, the dining car manager crosses the tracks and enters a tiny butchery. He emerges laden with juicy Patagonian steaks. Lunch is served soon after this stop as the locos are opened out to lift the train up the Rio Chico gorge.

The halfway point is reached at El Maiten. Although only a tiny desert outpost, El Maiten has a busy rail yard. The line's workshops and main loco shed are here. The hard worked Baldwins are replaced by equally old Henschel 2-8-2s. The knowledge that the nearest diesel is over 200 km away warms the heart.

Across the street from the station is Senor Jones's Hotel Vasconia. Despite his name, the proprietor speaks little Welsh and less English, but hospitality does not require a common language. Dinner is served beside the pot-bellied stove as a blizzard throws itself around the surrounding peaks. Snow flakes melt against the window; the roar of wheel slip and the clunk of draw gear briefly drown the sounds of the gale as a night freight struggles out of town on icy rails.

The blizzard continues through the night. By morning the wind has dropped, the clouds are clearing and the desert has been transformed. All is white. In contrast to the snow bound tranquility of the landscape, the shed is bustling as locos are hastily fitted with snow ploughs.

Even without snow El Maiten marks a change in lineside scenery. Locos reaching this point have struggled over a series of cliff-crowned ridges of eroded red rock. But now the line is much closer to the great Cordillera and is forced by geography to change direction to the south. Although the train still faces several heavy grades, much of the remainder of its journey takes it across wide desert vistas. Snow capped ranges parallel the line.

At Lepa, empty rails glinting in the afternoon sun approach from a massive wall of mountains. A faint haze rises from a distant valley. The haze becomes an eruption of smoke, rising blackly against the high snows. With a boom of oil fire and a rattle of carriages the train passes into the approaching dusk. The fertile valley of Esquel is still some hours ahead.

The abandoned station of La Cancha's only significance in summer is its water column. But the column must be protected from freezing in winter, so a fire is maintained at its base by a crew whose other duties include snow clearance. These intrepid souls spend the lonely winter months in the former stationmaster's cottage. Every couple of days the screech of brake blocks, the rush of water and chatter of passengers waiting to be on their way bring the place to life.

La Cancha's remoteness, tranquillity and obvious proximity to the railway occasionally attract dirt road toughened railfans. It is an ideal camp site. A group of trees, fed by decades of water tower overflow, provide firewood and shelter from the Patagonian wind. The water column has even been used as a shower, although this is not recommended in winter when the cry of the brass monkey has been heard.

At night the camp fire is a tiny splash of warmth beneath a canopy of stars. A rising moon bathes the Andean snows in eerie silver light. A far away whistle wails across the vastness of the night. It is some time before the impatient stammer of narrow gauge exhaust wafts in on a faint breeze. The train arrives, drinks and leaves — a wild creature in a wild land.

No railway can function without crews and the personalities of the men who work on this line make an extended visit a delight. Typical of El Maiten's enginemen is Ricardo Verona. Ricardo, with his trusty maté bowl, has the uncanny ability to be driving almost any loco that appears on the northern part of the line and appears to know exactly where the cameras are set up. The chimney invariably spews black smoke at just the right time. He loves his job: "Me gusta las machinas"; but his usually irrepressible good humour is dampened by rumours of the line's possible closure. The new president has pledged to reduce government spending and the Esquel line runs at a loss.

With luck, these rumours will come to nothing and Ricardo and his beloved "machinas" will continue to serve the lonely settlements of Patagonia. RKS

A snow covered spur of the Andes forms a spectacular backdrop to double Henschels on an Esquel bound passenger near Lepa.
April 1980 CS

▲ At Ojos de Agua. It was so cold that a fire had been lit to stop the water column from freezing. July 1989 RKS

We dare you to get the same shot with a diesel! Ojos de Agua. July 1989 MH ▶

▲ Baldwin No 22, in the capable hands of driver Ricardo Verona, accelerates away from the water stop at Mamuel Choique. July 1989 MH

◀ Some hundreds of metres higher, the same train breasts the summit of the grade between Mamuel Choique and Aguada Troncoso. July 1989 RKS

Sheep are not exactly the normal herd for macho gauchos. A pair of Baldwins occupy the background with a northbound passenger near El Maiten. October 1974 CS
▼

A constantly reinforced view of the Esquel line is that of a baby train in an adult landscape. Baldwin No 20 is hard at it near Ojos de Agua providing a balancing "passenger" working made up of goods and guards vans following a strike.

July 1985 RKS

On the footplate of No 22.
July 1989 MH

Driver Ricardo Verona is very fond of his maté, a herbal tea which is imbibed via a wooden bowled, metal stemmed pipe. He is also fond of cigarettes and a bewildering array of alcoholic beverages. But he is "addicted" to driving steam. Photographed in Cerro Mesa.
July 1989 RKS

Double Baldwins blast upgrade from Fitalancao.
July 1990 MT

Dawn at Mamuel Choique. The two Baldwins have watered, the passengers have stamped life back into frozen feet and the train is under way again.
June 1980 CS

2-8-2 No 22 receives attention at Mamuel Choique while working an Ingeniero Jacobacci to Cerro Mesa freight. July 1985 GB

The wild west with a Teutonic touch! A single Henschel rocks towards Lepa with a southbound freight.
June 1980 CS

LA VASCONIA
HOSTERIA
HELADOS

FERROCARRIL NACIONAL
109
GENERAL ROCA

THE PEOPLE

EL MAITEN
FERROCARRILES ARGENTINOS

TRENES

THE MACHINES

◀ Pre dawn at Ingeniero Jacobacci, the temperature was minus 10 degrees C and the passengers were all inside.
July 1989 SMcC

Carriage windows gleam with first light as a pair of Baldwins hustle the passenger towards Ojos de Agua. July 1990 SMcC

▼

The late running southbound passenger strikes dawn on the dual gauge near Ingeniero Jacobacci. The train runs on the narrower rail combination at right. July 1989 MH

Sun sliding down the flank of a glacial valley illuminates Henschel power south of El Maiten. July 1990 MH

Hills and horses.
A freight northbound from
El Maiten. June 1990 GB

A fresh crew and fresh engine. A Henschel takes its Esquel bound passenger at speed along the wide valley near El Maiten. July 1990 GB

Freshly outshopped Henschel No 104 poses for the camera at Cerro Mesa.
July 1989 MH

The waters of the Rio Chubut at El Maiten are never warm. Icy wet socks were forgotten, however, when the southbound passenger approached four hours late this evening. July 1989 RKS

◀ A large slice of the freight action on Esquel is haulage of oil tankers with fuel for the engines. Henschel No 104 is caught here on unfamiliar territory to the north of El Maiten. Rio Chico gorge. July 1989 MH

▲ Another Andean storm brews in the background as No 22 trundles its train towards Mamuel Choique.
July 1985 GB

◀ An S bend in the middle of the grade between Cerro Mesa and Aguada Troncoso provides the setting for a sunset contest between a Baldwin and maximum tonnage. September 1988 HS

VORDERNBERG

THE VORDERNBERG — EISENERZ RAILWAY

AUSTRIA

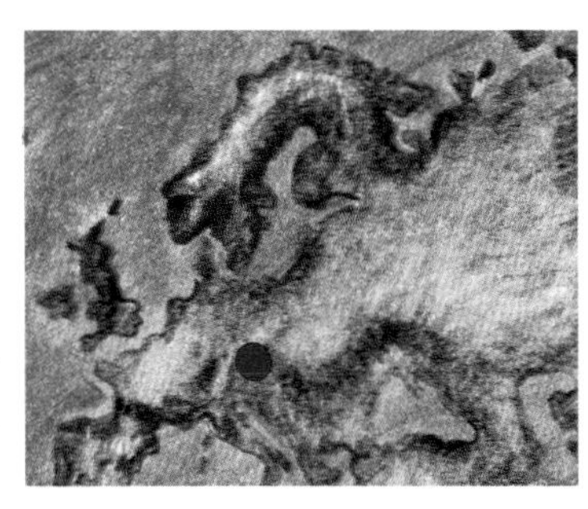

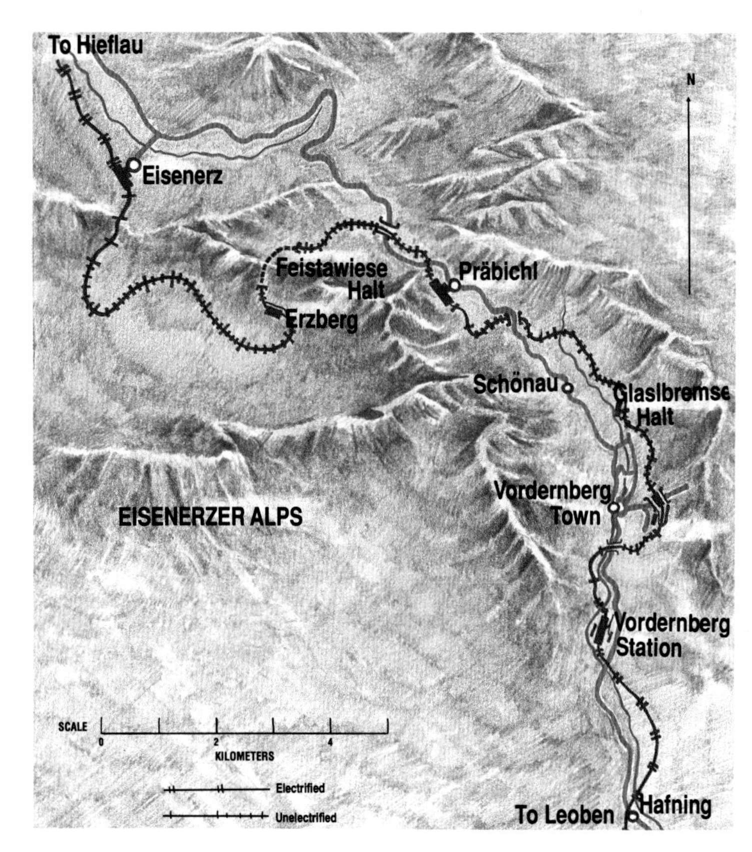

STATUS:	Single Track Secondary Connecting Line — ABT Rack
GAUGE:	4' 8 1/2" (143.5cm)
LENGTH OF LINE:	12 1/2 Miles (20 Kilometres)
RULING GRADE:	1 in 14
TRAFFIC:	Railbus and Steam Passenger Services Iron Ore
MOTIVE POWER:	97 Class 0-6-2T 1890 Austrian State Railways 197 Class 0-12-0T 1912 Austrian State Railways

Ore from the Iron Mountain

Nestled deep in a fold of the Austrian Hochschwab Range lies the village of Vordernberg, a cluster of square, utilitarian houses leavened with classic, thick-beamed traditional dwellings. The feet of the surrounding mountains are meadow-green in summer, their flanks are girt with the dark of pines and the highest peaks reflect the bright grey-brown of sheer rock.

The village is quaint in summer.

The winters in Vordernberg are long. Not content at crowning the heights, the snow claims the forest, pastures and houses with a deep white enthusiasm. The precincts resound to the hiss of toboggans, the stamp of boots and the jangling scrunch of snow chains. Ancient gasthofs beckon with the warmth of open fires, tales of the piste, hearty meals and the mellowing effect of strong drink.

The village comes into its own in winter.

This beguiling picture of Germanic bliss is repeated throughout the alpine valleys of west-central Europe and is not in itself exceptional. However Vordernberg once had an added zest . . . it had a railway. A rack railway. A steam rack railway!

The business of this railway was to haul mineral traffic from the Erzberg, an entire mountain of iron ore. To do so, the banked empty trains had to climb a long 1 in 14 grade to the summit of a mountain pass at a tiny hamlet called Präbichl and then to descend a couple of kilometres to the mine. Loaded workings had only to battle the short climb to the summit tunnel before dropping down the valley to the electric railhead below Vordernberg village. Up until the mid-70s, the railbus which provided a passenger link to the other electric railhead at Eisenerz was supplemented by a daily steam powered service.

The motive power at the time of these photos was a brace of 85 year old 97 Class 0-6-2 tanks, some of which were Giesl ejected and a sole surviving 1912-built 0-12-0T of Class 197, also Giesl fitted.

It has been said of a number of the world's more scenic lines that a careless photographer could fall over with a cocked camera, accidentally press the shutter button and still get a great shot. Vordernberg was such a place in winter. It guaranteed all and sundry the twin joys of steam and snow and occasionally rewarded patience with sun as well.

To gain a more complete appreciation of the Erzbergbahn, one could pay 25 schillings (inclusive of 16% tax) at the depot for an engine rider's pass to the mine and back. The trip up the line commenced on the flat in Vordernberg yard. No sooner had your engine settled into its rhythm than the driver started feeding steam into the compound inside cylinders to ensure a smooth transition onto the first rack section. To help him in this task, a small bullseye device mounted at the front of the water tank turned in time with the revolutions of the rack gear. Another short adhesion section and then it was onto the main grade, the only respite coming if a cross was to be made with a downhill working at Vordernberg Markt, the closest station to the village proper.

The prospect of spending several hours freezing on the exposed side of one's body, roasting on the firebox side, being deafened by clanging machinery and progressively covered in soot while blue-jacketed men toiled red-faced beside one in a cramped iron cab must seem to all sane people a stark reminder of the horrors of the industrial revolution. Yet, as the train rose up the valley, the rider was quickly bewitched by the scenery, the sight of the exertions of the engine at the other end and the total sensory experience of steam on the move. The enginemen were infinitely patient with questions posed in execrable German or shouted English and their gestures in the reverberating confines of the cab did much to convey what vocabulary could not.

All too soon, Präbichl came into sight through the pines. Then it was through the tunnel and down to the mine. Time spent in the slanting snow under the arc lights in the loading area of the mine allowed inhabitants of temperate climes to form a lurid vision of Siberia. The loaded trains were always split in two for the brief bunker-first thrash back to the summit. Once the full rake of wagons was reassembled, one could enjoy the comparative quiet as the train dropped down past the village with the rack gear acting as brake.

Handshakes all round at the end of the trip then back to the Gasthof Gruber to sample all the aforementioned delights. Between Helmut's drinks and his wife's cooking, the cold rapidly became a thing of the past. In the 1970s the visitors' book at the "Gruber" was more than a record of transient guests, it literally served as a postbox for information on the whereabouts of friends and for circulation of news on the wellbeing of the steam world.

Drop the word Vordernberg into the middle of any discussion involving international railway enthusiasts. See the smiles — Prosit! M. H.

Steam in monochrome. The first rack section out of Vordernberg yard.
February 1976 CS

Both the rack and adhesion gear on 97.207 are getting a solid workout in this Yuletide scene. December 1977 RKS ▶

The standard combination on the rack was 0-6-2Ts fore and aft. When the sole diesel was working it would operate alone, as would the 197 Class on most occasions. This train is approaching Vordernberg Markt. November 1974 MH ▼

197.303 leads a 97 Class with the normal half train length of loaded wagons through the drifting snows near Präbichl.
November 1974 RKS

The musclebound 197 Class 0-12-0T is going it alone with a string of empties between Vordernberg and Vordernberg Markt stations.
November 1974 MH

In a classic view of the Vordernberg rack line, two 97s, a 1 in 14 grade and nearly solid exhaust combine before the camera. December 1977 GB

38892

▲

Is he still doing his share back there? Hang on, I'll have a look! November 1974 RKS

◀ It was minus 4 degrees C, the icy road uphill behind the Gasthof Gruber had provoked a few spills in getting to the photo location and the train ran on time. After it passed a whiteout made the return to the Gruber even more interesting.
November 1974 RKS

▲

The track maintenance team are making a foray above Vordernberg Markt.
April 1975 MH

A room with a view. The Gasthof Gruber. April 1975 MH ▶

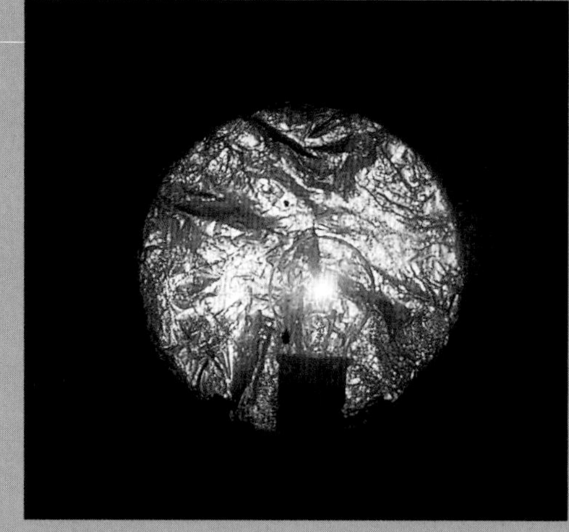

▲ Marker light, 97.209.
November 1974 MH

◀ The end of a day's skiing in Präbichl.
April 1975 GB

▲ The steepness of the grade is highlighted in this winter arrival at Vordernberg Markt.
February 1976 CS

◀ The silence at Präbichl is broken by the soft hiss of steam, the toot of the railbus and the thud of snow falling from branch to forest floor.
December 1977 RKS

Even the toughest tanks need a rest occasionally. 197.303 at the coaling point in Vordernberg yard.

May 1975 MH

After waiting for the arrival of the railbus, a pair of 0-6-2Ts propel their load back onto the main line at Vordernberg Markt station.

May 1975 MH

For some the Erzbergbahn was a near religious experience.
Above Vordernberg Markt.
November 1974 RKS

▲ The leading locomotive's exhalations had already cleared by the time the rear engine shuffled into view.
December 1977 RKS

◀ The climb out of Vordernberg yard was steep enough to justify rack assistance and short enough to cause the fireman only a moment's toil on 97.207. December 1977 RKS

A 97 leads the sole surviving 197 away from Vordernberg with empty ore wagons for the Erzberg. November 1974 MH

Uphill train. November 1974 MH

High above the village a 97 is silhouetted against the snow.
November 1974 MH

Winter wonderland. Yet another in the procession of banked trains climbs above Vordernberg village.
December 1974 MH

THE GARDEN ROUTE

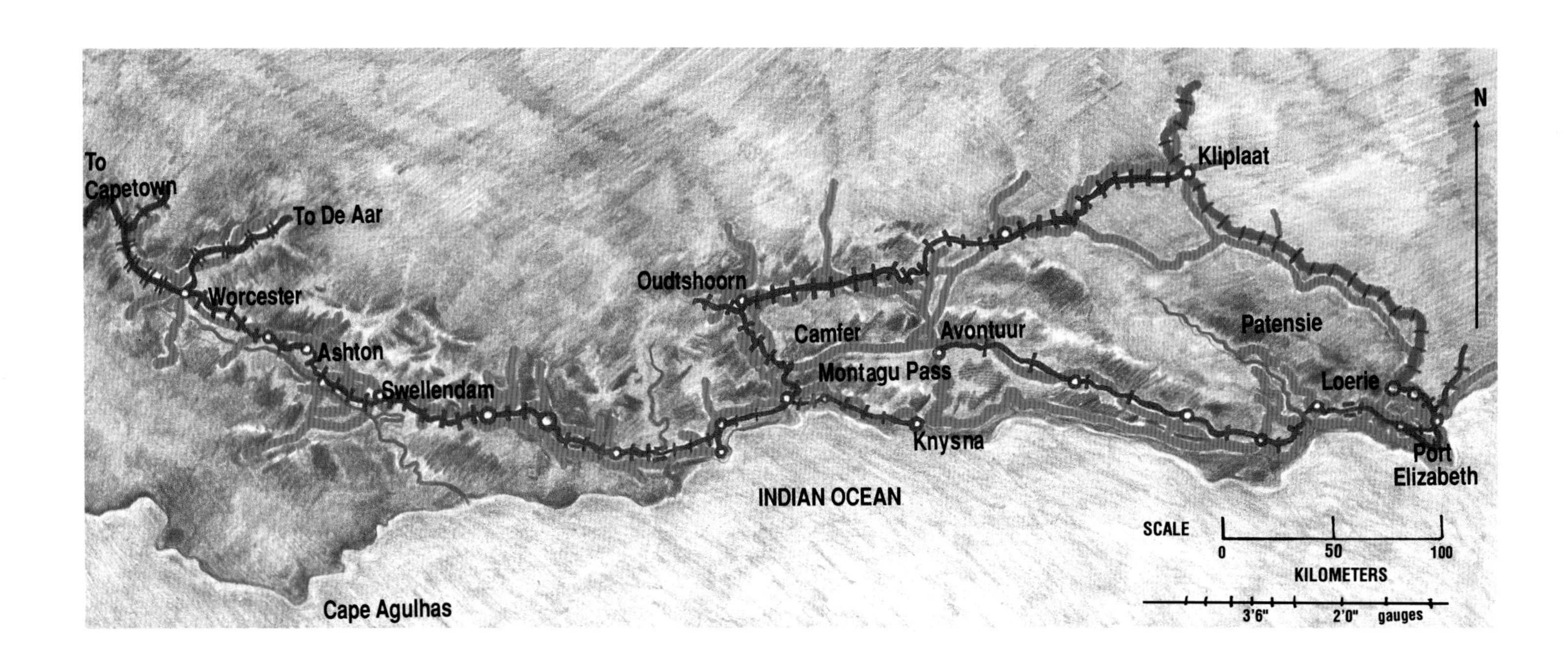

THE WORCESTER —PORT ELIZABETH & PORT ELIZABETH —AVONTUUR RAILWAYS, CAPE PROVINCE

SOUTH AFRICA

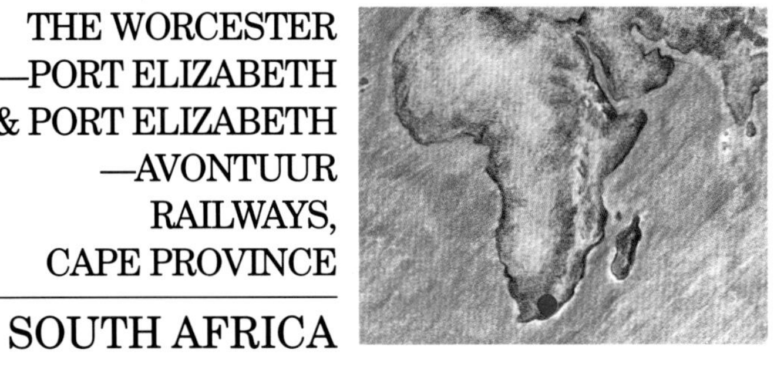

STATUS:	Single Track Secondary Main Line Single Track Branch Line (PE—AV)
GAUGE:	3' 6" (106.7cm) 2' (60.9cm) (PE—AV)
LENGTH OF LINE:	605 Miles (967 Kilometres) 178 Miles (285 Kilometres) (PE—AV)
TRAFFIC:	Six Days a Week Through Passenger, Suburban Passenger, General Freight Services General Freight Services (PE—AV)
MOTIVE POWER:	15AR Class 4-8-2 1914 North British* 16CR Class 4-6-2 1914 North British 19D Class 4-8-2 1937 Krupp* 24 Class 2-8-4 1949 North British GEA Class 4-8-2+2-8-4 1945 Beyer Peacock GMA Class 4-8-2+2-8-4 1952 Henschel* NG15 Class 2-8-2 1931 Henschel* NGG13 Class 2-6-2+2-6-2 Hanomag* * First Builder

The sounds of steam

Until 1979 the journey by rail along South Africa's Garden Route guaranteed the traveller an abundance of two things — stunning scenery and steam locomotives. The "Garden" is a web of lakes, forests and coastal lagoons, nestled between the mountains and the sea and stretching from Cape Town to Port Elizabeth.

No 9 Down, the Cape Town to Port Elizabeth passenger train, departed Cape Town Station daily except Saturdays and began its 1085 kilometre journey behind electric locomotives. At Worcester they came off and steam took over for the rest of the journey. It certainly wasn't the fastest train in the world, taking two nights and a day to reach its destination, but no-one seemed to mind as the route lay through some of the finest landscape in the Republic.

The Worcester to Riversdale section was a long standing stamping ground of the GMA Class Garratts or "Gammats" as they were known to their crews. They were not the most elegant of the many types of South African Garratt and the fact that drivers from Worcester had an obvious preference for running them backwards, with coal bunker leading, did not help. However, what they lacked in looks was more than compensated for by their performance in front of the jagged bulk of the Langeberg Mountains.

The Riversdale — Mossel Bay — Oudtshoorn section was worked by the older, hand-fired GEA Garratts until 1975 when GMAs, displaced from Natal by dieselisation, were transferred to Voorbaai Depot. In this section and at the foot of Montagu Pass lies the town of George, junction for the Knysna branch. A small shed houses the immaculately kept 24 Class 2-8-4s that still work this line. Happily the South African Railways recognises its potential as a tourist railway, running as it does through the heart of the "Garden".

Montagu Pass was a place of dreams for railway enthusiasts!

Above the quiet streets of George the Outeniquas rise as a beautiful backdrop. At night you could hear the sounds of steam. The Garratts on Oudtshoorn bound trains made a rapid departure in anticipation of the climb ahead. Eventually the characteristic stammer of their two engine units faded into the darkness, only to return the best part of an hour later, slower and quieter this time, as the train neared Power, halfway up the mountain.

A ride over the pass was always exciting. The line twists its way around the barren mountainsides, clinging to tiny ledges with drops of up to a thousand feet immediately below. The train wound its way upwards, seemingly forever, the wheel flanges screaming their protest at every curve and the mighty Garratt roaring away up front. The gradient is 1 in 36 and there are eight tunnels through solid rock, some unlined, enabling the railway to make its ascent.

On 9 Down, the dining car was a good place in which to sit back and enjoy the performance and spectacular scenery. Here aromas from the kitchen and coal smoke blended and wafted amongst the polished wood panels. Passengers lounged, their Lion Lagers forgotten, watching the contortions of the train whilst receiving first class service amid a setting of silver cutlery and white linen. From their position seven cars back, they were often amazed to see that the Garratt hauling the train was running almost parallel to their carriage — but in the opposite direction!

In GEA days, firemen needed nerves of steel to match their muscles. In hot weather, with the cab doors open, firing one of these machines was like walking a tightrope . . . and there was no safety net in the valley below. Some measure of relief finally came in steam's last years with the arrival of the mechanically stoked GMAs.

On arrival at Oudtshoorn the Garratts were replaced by conventional locomotives. Class 19D 4-8-2s were used on the run to Klipplaat at the other end of the Little Karoo desert. The Oudtshoorn 19Ds had torpedo tenders, giving increased water capacity for the run through this semi-arid, yet beautiful region. The inland vegetation here is in striking contrast with the lushness of the coast. Thorn trees, Karoo-bush and succulents dot the landscape and the purple hued Grootswartberge Ranges fill the horizon. From Klipplaat, the final sector to Port Elizabeth was worked by Class 15AR, an older and larger 4-8-2 type locomotive.

Until 1975 the last Pacifics in the land hauled 11 car locals from Port Elizabeth to Uitenhage. The 16CRs were worked to their limits on these trains. Short, high speed dashes between stations were required to keep the 'tables. The sounds of wide open regulators and long cutoffs were the norm everywhere on the system, but trainwatching from the road overbridge at the New Brighton end of Sydenham loco depot was a special experience. Drivers on Port Elizabeth-bound locals powered downgrade towards Sydenham Station. With speed approaching 100 kilometres per hour, many waited until underneath the bridge before shutting off in their usual flamboyant style. This final crescendo of stacktalk, with the Johnson bar pushed hard against the stop for a few seconds before closing the regulator, was truly thunderous.

These men really knew how to drive steam locos. Not only could they make them go, they had stopping them down to a fine art as well. In a swirl of brake dust, the 11 cars would pull up neatly in Sydenham platform. In all my time as a fireman at P.E. I did not see an overshoot anywhere on the system. But then if you were a suburban link driver you had a reputation to keep!

A large hoarding advertising cigarettes beside North End Station said it all: "Life Smaak Raak"—"Life Tastes Great". GB

The height of Cradock Peak provides a scenic counterpoint to the might of a hand fired GEA battling up Montagu Pass.

February 1974 RKS

Waiting for the train at Swellendam. September 1978 MH

The undulating green, mountain-backed path of the Worcester to Riversdale section had played host to GMA Class Garratts since 1958. No 4109, a member of that tribe, is therefore quite at home on the 02.40 Riversdale to Worcester freight near Voorhuis. March 1979 MT

In a scene which would have moved Van Gogh, GMA 4126 rises from a fold in the ground at the head of the 16.45 Riversdale — Mosselbaai passenger. Near Resiesboom siding. July 1976 MT

The line of a ridge near Oupad erupts in steam as yet another heavy goods train assaults the northern approach to Montagu Pass. August 1978 GB ▶

Sunrise at Camfer. The permanently attached water tank of GMA 4130 is filled to the brim for the climb up the northern side of Montagu Pass. April 1979 MT ▶

The shuffling roar of the labouring 4-8-2+2-8-4, the clang of shovel on firebox door and the squeal of flanges on rail reverberate over the pass as over 500 tons of train nears the cloud layer. March 1974 RKS

▼

GMA 4119 reflects the dying rays of a winter sun as it skirts the Langeberg Range near Voorhuis with the 13.15 Worcester — Riversdale mixed.
July 1976 MT

An immaculately turned-out GMA 4073 accelerates the P.E. — Cape Town passenger away from Oudtshoorn after taking over from a 19D Class 4-8-2.
April 1979 MT

Oudtshoorn locomotive depot under threatening skies.
August 1978 MT

First light at Middelplaas.
April 1979 MT

The lighting arrangements were anything but conventional at Dysseldorp when this torpedo tendered 19D paused with a Klipplaat bound train. July 1974 GB

A pencil thin shaft of sunlight squeezes in between mountaintop and lowering cloud to limn the sides of 19D 3368 at the point of the 15.45 P.E. to Cape Town passenger. Near Vanwykskraal.

August 1978 MT

Tootsie. June 1989 MH

An in bound working steps onto the Kaaimans River bridge at Wilderness.
June 1989 MH

The bird life is little disturbed
by the passing of a Knysna
bound train near Sedgfield.
September 1978 RKS

▲
A narrow gauge Garratt is hard at it on Loerie Bank with a load of Limestone.
October 1973 CS

Small loco, big thirst. Although on 2' gauge, the NGG 13/16 Class packed a punch which put some of the world's 3'6" gauge locomotives to shame. Loerie.
October 1973 CS
◀

Cab side, 2-8-2 No NG 117.
June 1989 MH

Loerie Bank was a test for every train, but this NG 15 certainly had its measure.
October 1973 CS

At rest at Humewood Road depot. **March 1974 RKS**

The sawtoothed skyline of the Kouga Mountains provide a dramatic backdrop for NG 15 No 145 on a load of bogie vans. Leaving Misgund.
April 1979 MT

Pacific No 825 awaits the witching hour of 23.05 for the start of its dash to Uitenhage. Port Elizabeth Station.
February 1975 GB

New Year's Eve at Port Elizabeth Station.
December 1974 GB

15AR 1807 and 16R 800 meet briefly at Uitenhage, while taking their turn on P.E. "subbies". October 1974 GB

Lowly duties indeed for a Pacific. This 1974 scene on the coal stage at Sydenham Shed would have featured an 11 Class 2-8-2 only months earlier. September 1974 CS

4-6-2 No 810 gets away from P.E. dockland with another 11 car commuter consist for Uitenhage. September 1974 GB

DARJEELING

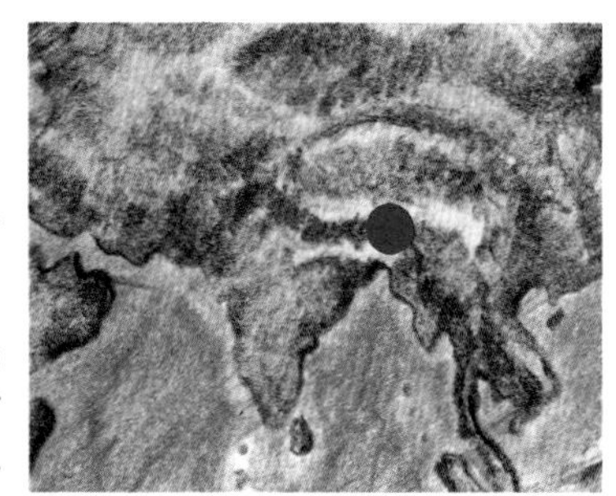

THE DARJEELING HIMALAYA RAILWAY, WEST BENGAL

INDIA

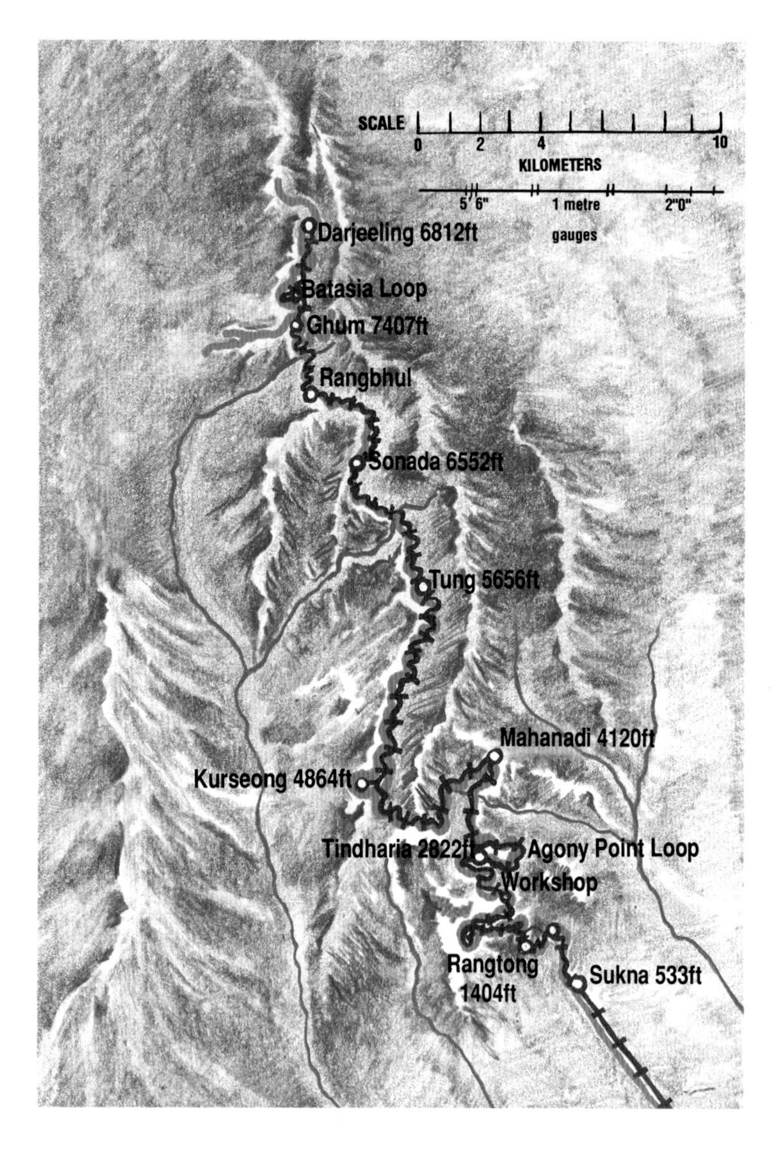

STATUS:	Single Track Branch Line
GAUGE:	2' (60.9cm)
LENGTH OF LINE:	54 Miles (87 Kilometres)
RULING GRADE:	1 in 20
TRAFFIC:	Through Passenger Services Local Passenger Services Freight Mainly For Railway Purposes
MOTIVE POWER:	B Class 0-4-0ST 1892 Sharp Stewart* * First Builder

Into thin air

The Darjeeling Himalaya Railway (DHR) is one of the wonders of the world. Although it climbs only into the Himalayan foothills, the DHR, known affectionately as the "Toy Train" to locals and tourists alike, is essentially a mountain railway. It ascends from 533 feet above sea level at Sukna, where the climb really starts, to 7406 feet at Ghum before dropping down to 6812 feet at its Darjeeling terminus.

The DHR is also one of the world's great train rides. After an overnight train trip from Calcutta and a brief encounter with Indian bureaucracy, the excitement begins. For over eight hours the passenger can enjoy the sight and sound of a diminutive 2 foot gauge loco attacking the grades with gusto. From New Jalpaiguri to Darjeeling is but 54 miles and a taxi or a bus is far quicker, but the DHR provides an unforgettable travel experience.

The lower section of the line passes through thick jungle, winding around huge banyan trees and crossing a number of ravines. Higher up the air is cooler and the vegetation less luxuriant. Every so often the terrain requires the train to resort to unusual means to gain extra height. Zig zags and spirals are both used. The middle leg of a zig zag commonly crosses the road, adding to the spice of the journey.

The DHR carriages are tiny and many unwary Westerners have cracked their heads on the door lintels. You don't really get into these carriages; rather, you put them on. The older carriages have ancient upholstered armchairs, while the newer variety rejoice in bigger windows. Whatever the weather, all windows should be opened to maximise the aural delights of uphill travel. On some cars the toilet window may also be opened and the enthroned passengers are thus able to continue their enjoyment of the scenery whilst in motion.

A troop of monkeys inhabits the woods below Kurseong and they range up and down the hillside with gay abandon. Goat-nibbled hydrangeas dot the sides of the sheer upper slopes, while the less steep areas are covered with tea bushes which grow right up to trackside at locations such as Margaret's Hope plantation.

The DHR follows the line of the Hill Cart Road very closely, frequently crossing from side to side to baffle unwary motorists. It also enables the train rider to see the various blue and white exhortations to motorists painted on the cliff faces. "Slow has four letters, so has life. Speed has five letters, so has death"; "Mind you left your child similing (sic) at home" and "If you drive like the hell you will get there soon", to quote but a few. The spelling varies from notice to notice and year to year.

Perched on the buffer-beam of every engine toiling uphill are two men, muffled up in greatcoats and scarves in winter, whose job it is to sprinkle sand on the tracks just in front of the leading wheels. These men are the world's most photographed railway employees. The rumour that they are all Spaniards named Manuel Sanders is not correct. Squatting on top of the coal in the bunker is another well-photographed railwayman equipped with a large hammer.

Nibelungen-like, he hammers the giant lumps of coal into smaller ones suitable for the miniscule firebox of a B Class 0-4-0ST. The driver and fireman round out the complement, which would be regarded as excessive on the largest broad gauge loco.

At Kurseong the train travels through the bazaar, fussing right past the shopfronts. A light-fingered passenger could easily purloin an orange or a banana, but would inevitably be confronted by an irate, rather breathless stallholder demanding payment. Dogs, goats and other animals seem to prefer the middle of the tracks for a snooze, reluctantly strolling off just as the train is about to fillet them.

School-age children pursue the train uphill, hurl themselves at the carriage sides and cling on while a creek or ravine is crossed, before dropping off to chase the train again. These junior Sherpas are amazingly fit. Despite the rarefied Himalayan air and their own deficiencies in the footwear department, they catch the train with ease and are seldom out of breath, despite keeping up the game for half a mile or more. Downhill racing doesn't seem to hold much attraction for them — perhaps it's too easy.

At Ghum, the DHR is set against a stunning panorama of snow covered mountains — the main Himalayan Range. Looming largest among the many nearby peaks visible in the clear, thin air is Kanchenjunga, the world's third highest mountain. Standing 28,168 feet tall, about 45 miles from Darjeeling as the vulture flies, Kanchenjunga's presence dwarfs the little railway. From the well known Batasia Loop both Kanchenjunga and the labouring DHR locomotives can be viewed simultaneously, weather permitting. However, the field of semi-frozen cabbages behind the old Ghum Monastery provides a far better vantage point. Kanchenjunga, the train and Batasia Loop all merge beautifully into a masterful composition. RB

Himalayan high rise. An uphill passenger train waters in the dead end platform at Kurseong between bouts with the 1 in 20 grades.

December 1983 RKS

MAYA

◀ An uphill train is almost buried in the foliage at the start of the climb. December 1986 MH

▲ The Kurseong to Darjeeling workers train emerges from a darkened cleft in the mountain side near Tung.
December 1983 GB

Late afternoon light catches a passenger service eight hours into its journey as it eases to a stand at the water tank at Tung.
December 1983 RKS

The light quality and colours of a Darjeeling day stretch Kodachrome to the limit. Below Tung. December 1983 RKS

Smoke hugs the hillside as a goods working wends its way uphill between Tung and Sonada. December 1983 RKS

Memories of a more leisurely age. Not a four wheel drive or bus in sight as bullock cart and steam engine engage in a 90 year old battle between Kurseong and Tung.

December 1983 RKS

With five zig zags and four spirals, the engineers who built the DHR were hard pressed to throw the line up more than 7000 feet from the plains. This train is on one of the major earthworks at Agony Point.

December 1986 RKS

A venerable B Class 0-4-0ST is nose on to the grade and skirting the pines.

December 1983 RKS

A recipe for fun. Take one hill, one billy cart and add children. December 1983 RKS

The Kurseong workers train winds through a village street beneath the bright remains of stranded kites. Between Tung and Sonada. December 1983 RKS

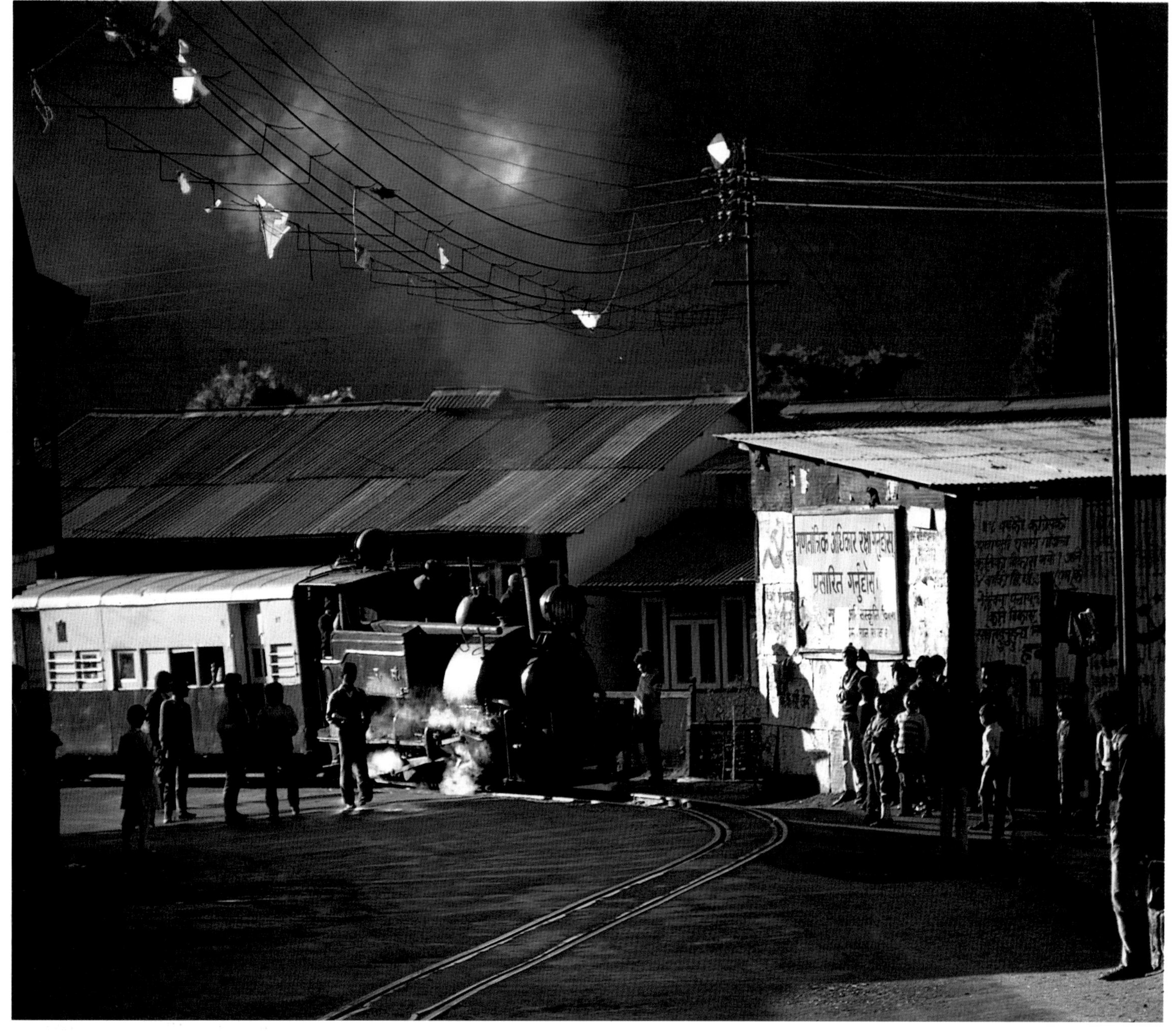

The hottest spot in town, Sonada at sunrise.
December 1983 GB

Man and machine will both benefit from the ministrations of the lighter-up who brings warmth to a freezing Darjeeling dawn. Locomotives are being prepared at the shed for the two morning passengers to New Jalpaiguri and a goods movement.
December 1983 GB

Waiting in the fog. Sonada.
January 1981 RKS

Sunset between Rangbhul and Ghum. December 1983 GB

Steam in silhouette, above Sonada. December 1983 GB

Steam in the ultimate setting. ▶▶ The train and intending passengers are dwarfed by the majesty of the main Himalayan Range as the 07.00 from Darj climbs Batasia Loop.
December 1983 GB

THE PEOPLE

THE MACHINES

Into thin air. Kanchenjunga looms in the distance as the 07.00 from Darjeeling arrives at Batasia Loop.
December 1983 RKS

A through train gets away
from the water stop at Tung
with the usual hangers-on.
You can almost taste the
atmosphere.

December 1983 RKS

The red stripe indicates the presence of mail, thick foliage cloaks the hill and the train snakes on. Between Rangbhul and Ghum. December 1983 GB

Sunset and the train is very near to the summit at Ghum.
December 1983 GB

THE CONTRIBUTORS

GEORGE BAMBERY
Photographer/ Journalist
— Melbourne, Australia
Story line — THE SOUNDS OF STEAM
Photographs — Front cover, 4/5, 42, 50, 51, 55, 63, 68/1, 81, 82, 87/1, 92, 100/1, 100/2 101, 103, 109, 116/1, 116/2, 118, 119, 120-121, 122/1, 123/4, 126, 127/1, 127/2.

ROBERT BELZER
Computer Systems Officer
— Melbourne, Australia
Story line — INTO THIN AIR
Photographs — 10/1, 12, 18/1, 19/2, 22, 23, 27/2, 29.

MALCOLM HOLDSWORTH
Taxation Officer
— Sydney, Australia
Story lines — A RIDE ON 19 NORTH
ORE FROM THE IRON MOUNTAIN
Photographs — 4/3, 4/4, 7, 10/2, 17/1, 17/2, 32, 34/2, 35/1, 38/1, 44/1, 44/2, 44/4, 44/5, 44/6, 44/7, 45/3, 45/4, 45/5, 45/6, 48, 49, 52/1, 52/2, 57, 58, 60, 62/2, 66, 67, 68/2, 70/1, 70/2, 76, 77, 78, 79, 84/1, 94, 95, 108, 122/2, 123/1, Back cover.

STEPHEN HOWARD
Carpenter
— Sydney, Australia
Photographs — 105, 123/2, 123/5.

ROBERT KINGSFORD-SMITH
Biologist
— Sydney, Australia
Story line — COLD DAYS, HOT ENGINES
Photographs — 14, 15, 16, 21/1, 21/2, 31, 34/1, 35/2, 37, 38/2, 44/3, 45/1, 45/2, 45/7, 53, 61, 62/1, 64, 65, 68/3, 71, 72-73, 74, 75, 83, 86, 96, 98/2, 106, 107, 110, 111, 112, 113, 114/1, 114/2, 115/1, 115/2, 117, 122/4, 124, 125, 128.

SHANE McCARTHY
Corporate Solicitor
— Melbourne, Australia
Photographs — 4/1, 8, 11, 13, 28, 46, 47, 122/5, 123/3.

SHANE O'NEIL
Customs Officer
— Katoomba, Australia
Photograph — 122/3.

COLIN SCHROEDER
Company Director
— Sydney, Australia
Photographs — 4/2, 9, 18/2, 20, 24-25, 31, 33, 36, 40-41, 43, 59, 69, 97/1, 97/2, 98/1, 102.

HOWARD STOKOE
Pharmacist
— London, United Kingdom
Photograph — 54

MICHAEL TYACK
Sales Manager
— Bristol, United Kingdom
Photographs — Frontispiece, 39, 84/2, 85, 87/1, 88-89, 90/1, 90/2, 91, 93, 99.

LAURIER WILLIAMS
Managing Director
— Sydney, Australia
Photographs — 19/1, 26, 27/1,